The ESSENTIALS of
ELECTRIC
CIRCUITS II

Staff of Research and Education Association,
Dr. M. Fogiel, Director

This book is a continuation of *"THE ESSENTIALS OF ELECTRIC CIRCUITS I"* and begins with Chapter 7. It covers the usual course outline of Electric Circuits II. Earlier/basic topics are covered in *"THE ESSENTIALS OF ELECTRIC CIRCUITS I"*.

Research and Education Association
61 Ethel Road West
Piscataway, New Jersey 08854

THE ESSENTIALS OF ELECTRIC CIRCUITS II

Printed in the United States of America

Library of Congress Catalog Card Number 87-61812

International Standard Book Number 0-87891-586-9

WHAT "THE ESSENTIALS" WILL DO FOR YOU

This book is a review and study guide. It is comprehensive and it is concise.

It helps in preparing for exams, in doing homework, and remains a handy reference source at all times.

It condenses the vast amount of detail characteristic of the subject matter and summarizes the **essentials** of the field.

It will thus save hours of study and preparation time.

The book provides quick access to the important facts, principles, theorems, concepts, and equations of the field.

Materials needed for exams, can be reviewed in summary form — eliminating the need to read and re-read many pages of textbook and class notes. The summaries will even tend to bring detail to mind that had been previously read or noted.

This "ESSENTIALS" book has been carefully prepared by educators and professionals and was subsequently reviewed by another group of editors to assure accuracy and maximum usefulness.

Dr. Max Fogiel
Program Director

CONTENTS

This book is a continuation of "*THE ESSENTIALS OF ELECTRIC CIRCUITS I*" and begins with Chapter 7. It covers the usual course outline of Electric Circuits II. Earlier/basic topics are covered in "*THE ESSENTIALS OF ELECTRIC CIRCUITS I*".

CHAPTER 7

POLYPHASE SYSTEMS

7.1 SINGLE-PHASE, 2-PHASE AND 3-PHASE SYSTEMS

7.1.1 SINGLE-PHASE, THREE-WIRE SYSTEM

The representation of a general single-phase, three-wire system is:

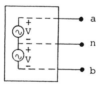

Since $v_{an} = v_{nb} = v$, $v_{ab} = 2v_{an} = 2v_{nb}$.

7.1.2 TWO-PHASE SYSTEM

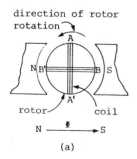

(a)

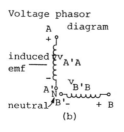

Voltage phasor diagram

(b)

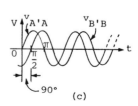

(c)

$$V_{BN} = V_{coil} \; \underline{/0^0}$$

$$V_{AN} = V_{coil} \; \underline{/90^0}$$

$$V_{AB} = V_{AN} + V_{NB} = V_{coil} \; \underline{/90^0} + V_{coil} \; \underline{/180^0} = \sqrt{2} \; V_{coil} \; \underline{/135^0}$$

7.1.3 THREE-PHASE SYSTEM

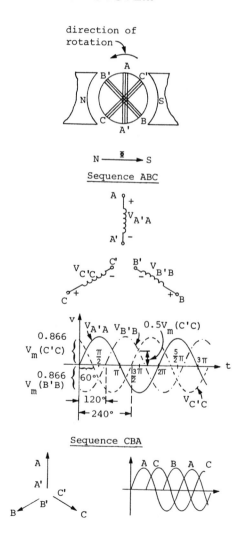

Characteristics

At any instant of time, the summation of all three phase voltages is zero, i.e.,

$$\Sigma \, (V_{A'A} + V_{B'B} + V_{C'C}) = 0.$$

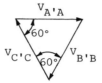

Note: In a three-phase system, the three coils on the rotor are placed 120^0 apart. (Assume each coil has an equal number of turns.)

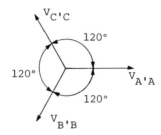

7.1.4 THREE-PHASE SYSTEM VOLTAGES

Sequence	(Note: V_L=line voltage)
ABC 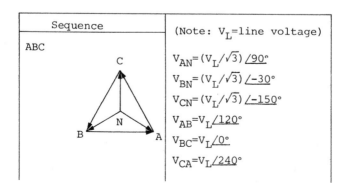	$V_{AN}= (V_L/\sqrt{3}) \underline{/90^\circ}$ $V_{BN}= (V_L/\sqrt{3}) \underline{/-30^\circ}$ $V_{CN}= (V_L/\sqrt{3}) \underline{/-150^\circ}$ $V_{AB}=V_L\underline{/120^\circ}$ $V_{BC}=V_L\underline{/0^\circ}$ $V_{CA}=V_L\underline{/240^\circ}$

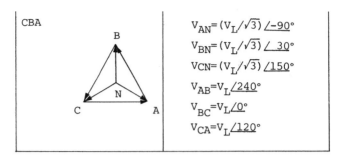

CBA	$V_{AN} = (V_L/\sqrt{3}) \underline{/-90°}$
	$V_{BN} = (V_L/\sqrt{3}) \underline{/\ 30°}$
	$V_{CN} = (V_L/\sqrt{3}) \underline{/150°}$
	$V_{AB} = V_L \underline{/240°}$
	$V_{BC} = V_L \underline{/0°}$
	$V_{CA} = V_L \underline{/120°}$

7.2 THE WYE (Y) AND DELTA (Δ) CONNECTIONS

7.2.1 WYE (Y) AND DELTA (Δ) ALTERNATORS

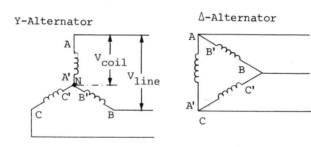

$$I_{coil} = I_{line} \qquad I_{coil} = \frac{1}{\sqrt{3}} I_{line}$$

$$V_{line} = \sqrt{3}\ V_{coil} \qquad V_{line} = V_{coil}$$

I_{coil} is more commonly referred to as I_{phase}.

7.2.2 THREE-PHASE Y-Y CONNECTION

Ideal voltage sources connected in Y (three-phase):

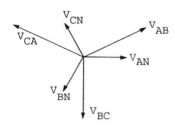

7.2.3 CHARACTERISTICS OF BALANCED THREE-PHASE SOURCES

1. $|V_{AN}| = |V_{BN}| = |V_{CN}|$ and $V_{AN} + V_{BN} + V_{CN} = 0$

2. If $V_{AN} = V_P \underline{/0^0}$ is the reference where V_P = rms is the magnitude of any of the phase voltages, then

$$V_{BN} = V_P \underline{/-120^0} \text{ and } V_{CN} = V_P \underline{/-240^0} \text{ (positive phase or sequence ABC),}$$

or

$$V_{BN} = V_P \underline{/120^0} \text{ and } V_{CN} = V_P \underline{/240^0} \text{ (negative phase or sequence CBA).}$$

Phasor diagrams of:

a positive sequence

$V_{CN} = V_P \underline{/-240^\circ}$

(V_P=phase voltage)

N

$V_{AN} = V_P \underline{/0^\circ}$

$V_{BN} = V_P \underline{/-120^\circ}$

a negative sequence

$V_{BN} = V_P \underline{/120^\circ}$

N

$V_{AN} = V_P \underline{/0^\circ}$

$V_{CN} = V_P \underline{/240^\circ}$

3. Phasor diagram of a line and phasor voltage relationship

V_{CN}

V_{CA}

V_{AB}

V_{AN}

V_{BN}

V_{BC}

$$V_{line} = \sqrt{3}\ V_{phase}$$

$$V_{AB} = \sqrt{3}\ V_P \quad \underline{/30^0}$$

$$V_{BC} = \sqrt{3}\ V_P \quad \underline{/-90^0}$$

$$V_{CA} = \sqrt{3}\ V_P \quad \underline{/-210^0}$$

7.2.4 DELTA CONNECTIONS

A balanced Δ-connected load with Y-connected source:

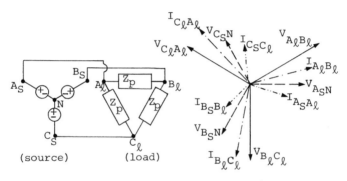

(source) (load)

Phasor Diagram

Given that $V_{phase} = |V_{A_SN}| = |V_{B_SN}| = |V_{C_SN}|$,

assume $V_{line} = |V_{A_SB_S}| = |V_{A_SC_S}| = |V_{C_SA_S}|$

where $V_L = \sqrt{3}\ V_A$ and $V_{A_SB_S} = \sqrt{3}\ V_{A_SN} \quad \underline{/30^0}.$

Then the phase currents are

$$I_{A_\ell B_\ell} = \frac{V_{A_SB_S}}{Z_p} \ , \quad I_{B_\ell C_\ell} = \frac{V_{B_SC_S}}{Z_p} \ \text{ and } \ I_{C_\ell A_\ell} = \frac{V_{C_SA_S}}{Z_p}$$

and the line currents are

$$I_{A_\ell B_\ell} - I_{C_\ell A_\ell} = I_{A_SA_\ell} \ , \text{ etc.}$$

Note: The three-phase currents are equal in magnitude,

i.e., $I_P = \left|I_{A_S B_S}\right| = \left|I_{B_S C_S}\right| = \left|I_{C_S A_S}\right|$,

$\qquad I_L = \left|I_{A_S A_\ell}\right| = \left|I_{B_S B_\ell}\right| = \left|I_{C_S C_\ell}\right|$

and $\quad I_L = \sqrt{3}\, I_P$.

7.3 POWER IN Y- AND Δ-CONNECTED LOADS

7.3.1 Y-CONNECTED LOAD

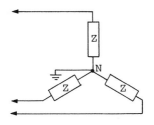

For a Y-connected load, the phase power with pf angle
$\theta = P_P = V_{phase} I_{line} \cos\theta$ (Note: $V_P I_P = V_P I_L = \dfrac{V_L I_L}{\sqrt{3}}$

The total power $= P_t = 3P_p$, or $P_t = \sqrt{3}\, V_L I_L \cos\theta$ where $V_L = \sqrt{3}\, V_P$.

Δ-connected load

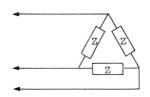

For a Δ-connected load, the phase power $= P_p =$ $V_L I_P \cos\theta$ (Note: $V_P I_P = V_L I_P = V_L \dfrac{I_L}{\sqrt{3}}$).

The total power $= P_t = 3P_p$ or $P_t = \sqrt{3}\, V_L I_L \cos\theta$.

Notice that the total power for any balanced three-phase load is equal to $\sqrt{3}\, V_L I_L \cos\theta$, where S_T (apparent power) $= \sqrt{3}\, V_L I_L$ and θ_T (reactive power) $= \sqrt{3}\, V_L I_L \sin\theta$.

CHAPTER 8

FREQUENCY DOMAIN ANALYSIS

8.1 COMPLEX FREQUENCIES

8.1.1 COMPLEX FREQUENCY

In general, the complex frequency s has the form $s = \delta + j\omega$ which describes an exponentially varying sinusoid. The complex frequency s consists of two parts:

1) The real part, δ, the neper frequency in nepers/sec.

2) The imaginary part, $j\omega$, where ω is   the   radian frequency in radians/sec.

The real part is related to the exponential variation; the imaginary part is related to the sinusoidal variation.

In general, a function $f(t)$ can be expressed in terms of the complex frequency s as

$$f(t) = k \, e^{st}$$

where k is a complex constant.

The characteristics of the function f(t) relate to the complex frequency s as follows:

s	f(t)
+	increases
-	decreases as t increases
0	constant sinusoidal amplitude

Note: Increasing the magnitude of the real part of s will increase the rate of the exponential increase or decrease. Increasing the magnitude of the imaginary part of s will increase the time function changing rate.

8.2 COMPLEX FREQUENCY IMPEDANCES AND ADMITTANCES

Table of complex frequency impedances and admittances for elements R, L and C:

element	impedance Z(s)	Admittance Y(s)
R	R	$\dfrac{1}{R}$
L	sL	1/sL
C	1/sC	sC

Note: $Z(s) = \dfrac{V}{I} = \dfrac{1}{Y(s)}$

8.3 THE S-PLANE (COMPLEX-FREQUENCY PLANE)

8.3.1 THE S-PLANE

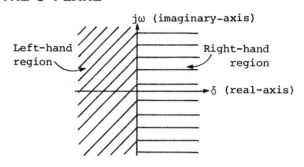

1) A point at the origin.──►corresponds to a DC quantity

2) Points on δ -axis ──► a) $\delta > 0$ ~exponential functions decaying

   b) $\delta < 0$ ~exponential functions increasing

3) Points on $j\omega$-axis ──► purely sinusoidal functions.

4) Points in ▨ (left-hand-region) ──► describe frequencies of exponentially-decreasing sinusoids.

5) Points in ▤ (right-hand-region) ──► describe frequencies of exponentially-increasing sinusoids. (i.e., frequencies of positive real parts,time-domain quantities).

8.4 POLES AND ZEROS

Consider a rational function of the form

$$H(s) = \frac{b_m s^m + b_{m-1} s^{m-1} +...+ b_1 s + b_0}{a_n s^n + a_{n-1} s^{n-1} +...+ a_1 s + a_0} .$$

It can be expressed as

$$H(s) = k \frac{(s-z_1)\ldots(s-z_m)}{(s-p_1)\ldots(s-p_n)}$$

where the zeros of $H(s)$ (i.e., $z_1, z_2, \ldots, z_m$) can be obtained by setting the numerator of $H(s)$ equal to zero.

The poles of $H(s)$ (i.e., $p_1, p_2, \ldots, p_n$) can be obtained by setting the denominator of $H(s)$ equal to zero.

Poles and zeros are indicated in the S-plane (complex-frequency plane) as follows:

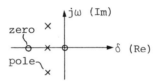

Procedures for graphical determination of magnitude and angular variation of frequency-domain function

Step 1: Find all poles and zeros of the frequency-domain function. Indicate them in the S-plane and, for the function to be determined, assign a test point on the S-plane corresponding to the frequency.

Step 2: Draw a corresponding arrow from each pole and zero to the test point.

Step 3: Calculate the length and angle of each pole arrow and zero arrow.

Step 4: Determine the magnitude of the frequency-domain function for the assumed frequency of the test point by the following ratio:

product of the zero-arrow lengths
product of the pole-arrow lengths

Step 5: Finally, use the formula

[Sum of zero-arrow angles]-[sum of pole-arrow angles]

69

to obtain the angular variation of the frequency-domain function evaluated at the test point.

8.5 RESONANCE (SERIES AND PARALLEL)

8.5.1 RESONANCE

In a network, when the voltage and the current at the input terminals are in phase, the network is in resonance.

Note: In resonance, power factor (pf) is unity.

8.5.2 PARALLEL RESONANCE

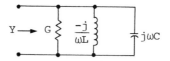

Characteristics

1) The complex admittance Y is

$$Y = G + j (\omega C - 1/\omega L).$$

2) Resonance condition for the circuit:

$$\omega C - \frac{1}{\omega L} = 0$$

$$\omega C = \frac{1}{\omega L}$$

$$\omega = \frac{1}{\sqrt{LC}} = \omega_0$$

where ω_0 = resonant frequency

or $f_0 = \dfrac{1}{2 \pi \sqrt{LC}} \dfrac{\text{cycles}}{\text{sec}}$

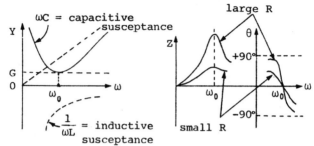

(a) Parallel circuit Y as a function of ω.

(b) Z and θ as a function of ω.

Note: At $\omega < \omega_0$, inductive susceptance > capacitive susceptance and Y is negative.

$\omega > \omega_0$, $\measuredangle$ Z is negative.

$\omega \to 0$, $\measuredangle$ Z is $+90^0$

4) Pole-zero representation of Y

$$Y(s) = K \frac{(s+\alpha - j\omega\alpha)(s+\alpha - j\omega\alpha)}{s}$$

where α = exponential damping ratio = $\dfrac{1}{2RC}$,

ω_d = natural resonant frequency = $\sqrt{\omega_0^2 - \alpha^2}$

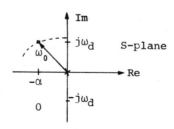

8.5.3 SERIES RESONANCE

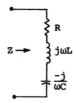

Characteristics

1) The complex impedance Z is

$$Z = R + j(\omega L - \frac{1}{\omega C}).$$

2) Resonance condition for the circuit:

$$\omega L - \frac{1}{\omega C} = 0$$

or

$$\omega = \frac{1}{\sqrt{LC}} = \omega_0$$

Hence, the resonant frequency $(f_0) = \dfrac{1}{2\pi \sqrt{LC}} \quad \dfrac{\text{cycles}}{\text{sec}}$

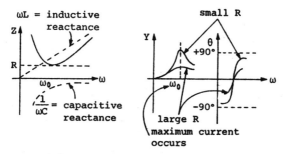

Z, Y, and θ as a function of ω.

When $\omega < \omega_0$, capacitive reactance > inductive reactance
and ∡ Z is negative.

When $\omega > \omega_0$, inductive reactance > capacitive reactance
and ∡ Z is positive approaching +90⁰.

When $\omega \to 0$, ∡ Z is -90⁰.

8.6 QUALITY FACTOR - Q

8.6.1 QUALITY FACTOR

By definition, the quality power Q is expressed as

$$Q = 2\pi \times \frac{\text{maximum energy stored}}{\text{total energy lost or dissipated per period}}$$

8.6.2 QUALITY FACTOR Q FOR SOME GENERAL CIRCUITS

CIRCUIT	QUALITY FACTOR, Q
1. $\quad$ RC series	$Q = 2\pi \times \dfrac{\frac{1}{2}I^2_{max}/\omega^2 C}{(I^2_{max}/2)\times R \times T} = \dfrac{1}{\omega CR}$
2. $\quad$ RL series	$Q = 2\pi \times \dfrac{\frac{1}{2}LI^2_{max}}{(I^2_{max}/2)\times R \times T} = \dfrac{2\pi f L}{R} = \dfrac{\omega L}{R}$ $T = \dfrac{1}{f} = \dfrac{2\pi}{\omega}$
3. $\quad$ RLC series	At resonance: $Q_0 = \dfrac{\omega_0 L}{R} = \dfrac{1}{\omega_0 CR}$ or $Q_0 = \dfrac{\omega_0}{\omega_2 - \omega_1} = \dfrac{f_0}{f_2 - f_1} = \dfrac{f_0}{BW\,(\text{Bandwidth})}$ Note: $\frac{1}{2}CV^2_{max} = \frac{1}{2}LI^2_{max}$ (at maximum)
4. $\quad$ RLC parallel	At resonance: $Q_0 = \dfrac{R}{\omega_0 L} = \omega_0 CR$

73

Note: Both circuits in (3) and (4) store a constant amount of energy at resonance.

8.7 SCALING

The method of scaling is used to ease the numerical calculations during networks analysis. There are basically two types of scaling: magnitude scaling and frequency scaling.

Magnitude scaling

A factor of K_m is increased for the impedance (z) of a two-terminal network with the frequency remaining constant, i.e.,

$R \rightarrow K_m R$,

$C \rightarrow C/K_m$ and

$L \rightarrow K_m L$.

Frequency scaling

A factor of K_f is increased for the frequency at any impedance, i.e.,

$R \rightarrow R$,

$C \rightarrow C/K_f$ and

$L \rightarrow L/K_f$.

CHAPTER 9

STATE - VARIABLE ANALYSIS

9.1 STATE - VARIABLE METHOD

Given a circuit with energy-storing elements (i.e., a capacitor and inductor), the circuit can be analyzed by using the state-variable method. A hybrid set of variables is selected (including capacitor voltages and inductor currents) to describe the energy state of the system.

By using a set of state variables, a set of n first-order, simultaneous differential equations can be obtained from the given nth-order differential equations (i.e., state equations).

9.1.1 CONDITIONS FOR THE STATE-VARIABLE METHOD

1) The state equation must be expressed in normal form, i.e., the derivative of each state variable must be expressed in terms of a linear combination of all the state variables and forcing functions.

2) The equations describing the derivatives must be of the same order as the state variables appearing in each equation.

9.2 STATE EQUATIONS FOR n-th ORDER CIRCUITS

9.2.1 STATE EQUATIONS FOR THE FIRST AND SECOND ORDER CIRCUITS

1) First-order circuit:

<u>State equation</u>

$$\dot{x}(t) = \frac{1}{RC} [v(t)-x(t)]$$

1) $x(t)$ represents the state variable $v_c(t)$.

2) $v(t) = Ri(t) + v_c(t)$,

$$\uparrow$$
$$i(t) = ic(t) = C \frac{dv_c}{dt}$$

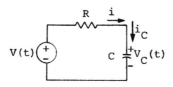

2) Second-order circuit

$$\dot{x}_1(t) = -\frac{R}{L} x_1(t) - \frac{1}{L} x_2(t) + \frac{1}{L} v(t)$$

$$\dot{x}_2(t) = \frac{1}{C} x_1(t)$$

$$x_1(t) = i_L(t)$$

$$x_2(t) = v_c(t)$$

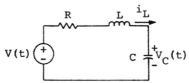

9.2.2 GENERAL STEPS TO OBTAIN THE STATE EQUATIONS FOR A LINEAR TIME-INVARIANT NETWORK

General steps to obtain the state equations for a linear time-invariant network

Method 1

Step 1: Assign state variables for the voltage across each capacitor and for the current through each inductor.

Step 2: Apply KVL and KCL to obtain a set of linear independent equations for each capacitor and inductor.

Step 3: Rearrange the equations obtained in step 2 so that all other variables in the network are in terms of the chosen state variables.

Step 4: Consider all the equations obtained in step 2 and 3. Simplify them so that the equations are expressed only in terms of the state variables and their corresponding derivatives. Therefore, all network variables not chosen as state variables are eliminated.

Step 5: Rearrange the equations obtained in step 4 in the compact form

$$\dot{x}(t) = Ax(t) + bu(t) \text{ (the normal-form equation)},$$

where $\dot{x}(t)$ = derivative of $x(t) = \dfrac{dx(t)}{dt}$,

$x(t)$ = n-column vector of all the state variables $(x_1, x_2, \ldots x_n)$ chosen in step 1 (i.e., $x(t) \triangleq [x_1(t) + x_2(t) + \ldots + x_n(t)]^T$,

A = constant $n \times n$ square matrix = system matrix (i.e., $A \triangleq (a_{ij})$), b = an n-column vector, and bu(t) = the forcing function vector due to the independent sources (i.e., $b \triangleq [b_1, b_2, \ldots, b_n]$).

Therefore,

$$
\begin{bmatrix} \dot{x}_1(t) \\ \dot{x}_2(t) \\ \vdots \\ \dot{x}_n(t) \end{bmatrix} = \begin{bmatrix} A_{11} & A_{12} \ldots A_{1n} \\ \vdots & \vdots \\ \vdots & \vdots \\ \vdots & \vdots \\ A_{n1} & \ldots \ldots A_{nn} \end{bmatrix} \begin{bmatrix} x_1(t) \\ x_2(t) \\ \vdots \\ x_n(t) \end{bmatrix} + \begin{bmatrix} b_{11} \\ b_{21} \\ \vdots \\ b_{n1} \end{bmatrix} u(t)
$$

9.3 NORMAL - FORM EQUATIONS

9.3.1 GENERAL PROCEDURES TO OBTAIN A SET OF NORMAL-FORM EQUATIONS

Method 2

Step 1: Obtain a normal tree for the given network using the nodal analysis as outlined in Chapter 5.

Step 2: Assign state variables for the voltage across each capacitor and the current through each inductor correspondingly, i.e.,

or, for the resistive tree branches or links, indicated by using a new voltage or current variable (if necessary).

Step 3: a) For each capacitor, apply KCL (as outlined in Chapter 2) to write a set of equations.

b) For each inductor, repeat part (a) but use KVL instead.

c) If any new voltage and current variables were assigned to the resistors, write the equation for R by using KCL and KVL. Then express v_R and i_R in terms of the state variables and source quantities. Otherwise, skip this step.

Step 4: Put all the equations obtained in step 3 in order and rearrange to obtain the normal-form equations.

9.4 STATE TRANSITION MATRIX-e^{At}

9.4.1 STATE TRANSITION - e^{At}

Let us represent the state equations in the normal form

$$\dot{x}(t) = Ax(t) + bu(t) \quad (At \ t=t_0, \ x(t_0) = x_0.) \qquad (1)$$

The solution of the matrix state equation (in (1)) is given by

$$x(t) = e^{A(t-t_0)}x_0 + \int_{t_0}^{t} e^{A(t-\tau)}bu(\tau)d\tau,$$

where $e^{At} \triangleq$ the state transition matrix, which describes the change of state of the system from zero to the state at time t where $e^{A(t-t_0)}$ is e^{At} evaluated at $t = t - t_0$.
(Note: $x(t)$ and $e^{A(t-t_0)}bu(\tau)$ are n-column vectors.)

To determine the state transition matrix e^{At}, the Cayley-Hamilton theorem is applied.

9.4.2 CAYLEY-HAMILTON THEOREM

1) By definition,

$$e^{At} \overset{\Delta}{=} \mu_0(t)I + \mu_1(t)A + \mu_2(t)A^2 + \ldots + \mu_{n-1}(t)A^{n-1} \qquad (2)$$

where A is an $n \times n$ square matrix and $\mu_0(t) \ldots \mu_n(t)$ are scalar functions of time.

2) For equation (2) to hold, the following conditions must be satisfied:

a) I = unity matrix.

b) The characteristic equation of the matrix A equals

   $\text{Det}[A - sI] = 0$,

where Si, $i = 1,2,\ldots,n$ of A are the roots of the characteristic nth-order polynomial equation and are called the eigenvalues of A.

9.4.3 GENERAL PROCEDURES TO OBTAIN THE STATE TRANSITION MATRIX e^{At}, GIVEN MATRIX A

Step 1: Obtain a matrix in the form $A - sI$.

Step 2: Equate: $\text{Det}[A - sI] = 0$ and solve for the roots (i.e., Si, $i = 1,2,\ldots,n$) of the characteristic equation.

Step 3: Express each root in n equations of the form $e^{tS_i} = \mu_0 + \mu_1 Si + \ldots + \mu_{n-1} Si^{n-1}$, and solve the scalar time functions: $\mu_0,\ldots,\mu_{n-1}$.

Step 4: Obtain the state transition matrix by substituting the time functions obtained in step 3 into Equation (2).

CHAPTER 10

FOURIER ANALYSIS

10.1 TRIGONOMETRIC FOURIER SERIES

10.1.1 DIRICHLET CONDITIONS FOR THE EXISTENCE OF A FOURIER SERIES

If $f(t)$ is a bounded periodic function of period T (i.e., $f(t+T) = f(t)$) and if $f(t)$ satisfies these Dirichlet conditions:

1) $f(t)$, if discontinuous, has a finite number of discontinuities in any period T;

2) $f(t)$ has a finite average value over period T;

3) The number of maxima and minima of $f(t)$ in any period T is finite

then $f(t)$ may be represented by a trigonometric Fourier series as described below:

10.1.2 TRIGONOMETRIC FORM OF A FOURIER SERIES

1) The function $f(t)$ is expressed over any interval $(t_0, t_0 + 2\pi/\omega_0)$ as

$$f(t) = a_0 + a_1\cos \omega_0 t + a_2\cos 2\omega_0 t + \ldots + b_1\sin\omega_0 t + b_2 \sin 2\omega_0 t + \ldots$$

$$= a_0 + \sum_{n=1}^{\infty} (a_n \cos n\omega_0 t + b_n \sin n\omega_0 t) \quad (t_0 < t < t_0 + 2\pi/\omega_0)$$

where ω_0 = fundamental frequency = $2\pi/T$

and $a_0 = \frac{1}{T} \int_{t_0}^{(t_0+T)} f(t)dt$ (where t_0 is assumed to be zero generally)

$$a_n = \frac{2}{T} \int_{t_0}^{(t_0+T)} f(t)\cos n\omega_0 t \, dt$$

$$b_n = \frac{2}{T} \int_{t_0}^{(t_0+T)} f(t)\sin n\omega_0 t \, dt$$

10.1.3 SPECIAL INTEGRATION PROPERTIES FOR SINE AND COSINE

1) $\int_0^T \sin^2 n\omega_0 t \, dt = \frac{T}{2}$

$\int_0^T \cos^2 n\omega_0 t \, dt = \frac{T}{2}$

2) $\int_0^T \sin n\omega_0 \, dt = \int_0^T \cos n\omega_0 t \, dt = 0$

3) $\int_0^T \sin k\omega_0 t \cos n\omega_0 t \, dt = \int_0^T \sin k\omega_0 t \sin n\omega_0 t \, dt$

$= \int_0^T \cos k\omega_0 t \cos n\omega_0 t = 0$

for $k \neq n$.

10.2 EXPONENTIAL FOURIER SERIES

A given function $g(t)$ can be expressed as a linear combination of exponential functions over the period t_0, $t_0 + 2\pi/\omega_0$, as follows:

$$g(t) = \ldots + G_{-n}e^{-jn\omega_0 t} + \ldots + G_{-2}e^{-j2\omega_0 t} + G_{-1}e^{-j\omega_0 t}$$

$$+ G_0 + G_1 e^{j\omega_0 t} + G_2 e^{j2\omega_0 t} + \ldots + G_n e^{jn\omega_0 t} + \ldots$$

$$= \sum_{n=-\infty}^{\infty} G_n e^{jn\omega_0 t} \quad (t_0 < t < t_0 + 2\pi/\omega_0),$$

it is assumed that $t_0 = 0$.

(Note: $T = 2\pi/\omega_0$.)

Therefore,

$$G_n = \frac{1}{T} \int_0^T g(t)e^{-jn\omega_0 t}\, dt$$

and

$$G_0 = \frac{1}{T} \int_0^T g(t)\, dt$$

Relationships between trigonometric and exponential Fourier series

Trigonometric series		Exponential series
a_0	$=$	G_0
a_n	$=$	$G_n + G_{-n}$
b_n	$=$	$j(G_n - G_{-n})$
$\frac{1}{2}(a_n - jb_n)$	$=$	G_n

10.3 COMPLEX FORM OF A FOURIER SERIES

The complex form of a Fourier series is given as

$$g(t) = \sum_{n=-\infty}^{\infty} C_n e^{jn\omega_0 t}$$

where C_0 is a complex constant $= \dfrac{1}{T} \displaystyle\int_{-\frac{T}{2}}^{\frac{T}{2}} g(t)e^{-jn\omega_0 t} \, dt$,

for $n = 0, \pm 1, \pm 2, \pm 3, \ldots$.

Note: $|C_n| = |C_{-n}|$, since $C_{-n} = C_n{}^*$.

10.3.1 SPECIAL CASE

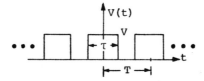

Given a train of rectangular pulses with period T, i.e.,

$$v(t) = \begin{cases} v & -\tau/2 < t < \tau/2 \\ 0 & \tau/2 < t < T - \tau/2, \end{cases}$$

then

$$C_n = \frac{1}{T} \int_{-\frac{\tau}{2}}^{\frac{\tau}{2}} A \, e^{-jn\omega_0 t} \, dt$$

$$= \frac{2A}{n\omega_0 T} \frac{(e^{jn\omega_0 \tau/2} - e^{-jn\omega_0 \tau/2})}{2j}$$

$$= \frac{A\tau}{T} \left[\frac{\sin(\tfrac{1}{2}n\omega_0 \tau)}{(\tfrac{1}{2}n\omega_0 \tau)} \right] = \frac{A\tau}{T} \, \text{Sa}(\tfrac{1}{2}n\omega_0 \tau),$$

where $\frac{\sin x}{x}$ = sampling function = Sa(x).

Thus, $v(t) = \frac{A\tau}{T} \sum_{n=-\infty}^{\infty} Sa(\tfrac{1}{2}n\omega_0\tau)e^{jn\omega_0 t}$.

10.4 WAVEFORM SYMMETRY PROPERTIES

Waveform symmetry	Properties
1) Even symmetry (i.e., cosine function)	a) $g(t) = g(-t)$ b) $b_n = 0$ c) $a_n = \frac{4}{T} \int_0^{T/2} g(t)\cos n\omega_0 t\, dt$
2) Odd symmetry (i.e., sine function)	a) $g(t) = -g(-t)$ b) $a_0 = a_n = 0$ c) $b_n = \frac{4}{T} \int_0^{T/2} g(t)\sin n\omega_0 t\, dt$
3) Half wave symmetry	a) $g(t) = -g(t+T/2)$ where T = period b) For n = odd, $a_n =$ $\frac{4}{T} \int_0^{T/2} g(t)\cos n\omega_0 t\, dt$

Waveform symmetry	Properties
3) Half wave symmetry	$b_n = \dfrac{4}{T} \displaystyle\int_0^{T/2} g(t)\sin n\omega_0 t\, dt$ $n = \text{even}, \quad a_n = b_n = 0$
4) Half wave and even symmetry	a) For $n = \text{odd}, \quad a_n =$ $\dfrac{8}{T} \displaystyle\int_0^{T/4} g(t)\cos n\omega_0\, t\, dt$ $b_n = 0, \; n = \text{even}, \; a_n = b_n = 0$
5) Half wave and odd symmetry	a) For $n = \text{odd}, \; a_n = 0$ and $b_n = \dfrac{8}{T} \displaystyle\int_0^{T/4} g(t)\sin n\omega_0 t\, dt$ $n = \text{even}, \; b_n = a_n = 0$

10.5 THE FOURIER TRANSFORM

By definition, the Fourier transform of $g(t)$ is

$$F\{g(t)\} = G(\omega) = \int_{-\infty}^{\infty} g(t)e^{-j\omega t}\, dt$$

The inverse Fourier transform of $F(\omega)$ is

$$F^{-1}\{G(\omega)\} = g(t) = \frac{1}{2\pi} \int_{-\infty}^{\infty} G(\omega)e^{j\omega t}\, d\omega,$$

as long as $\displaystyle\int_{-\infty}^{\infty} |f(t)| \, dt$ converges (i.e., $\displaystyle\int_{-\infty}^{\infty} |f(t)| \, dt < \infty$).

Thus, g(t) and G(ω) are called the Fourier transform pair.

Note: The Fourier transform of g(t) can also be expressed in terms of sine and cosine by using the Euler's identity:

$$e^{-j\omega t} = \cos \omega t - j\sin \omega t.$$

Hence,

$$G(\omega) = \overbrace{\int_{-\infty}^{\infty} g(t)\cos\omega t \, dt}^{R(\omega)} - j \overbrace{\int_{-\infty}^{\infty} g(t)\sin\omega t \, dt}^{I(\omega)}$$

$$= |G(\omega)| \underline{\bigg/ \theta(\omega)}$$

where

$$|G(\omega)| = [R^2(\omega)+I^2(\omega)]^{\frac{1}{2}}$$

and

$$\theta(\omega) = \tan^{-1}\left[\frac{I(\omega)}{R(\omega)}\right]$$

10.5.1 SOME USEFUL FOURIER TRANSFORM PAIRS

$$g(t) \iff G(\omega) = F\{g(t)\}$$

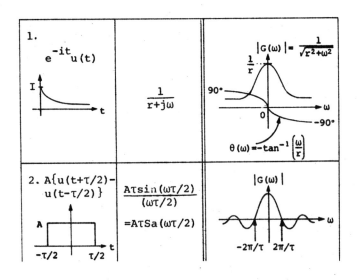

87

3. $\delta(t)$![graph]	1	$G(\omega)$ 1
4. 1 1	$2\pi\delta(\omega)$	$2\pi\delta(\omega)$
5. $u(t)$	$\pi\delta(\omega) + \dfrac{1}{j\omega}$	$\lvert G(\omega)\rvert$
6. $\operatorname{sgn}(t)$ 1 -1	$\dfrac{2}{j\omega}$	$\lvert G(\omega)\rvert$
7. $\cos\omega_0 t$ 1 -1	$\pi\{\delta(\omega-\omega_0) + \delta(\omega+\omega_0)\}$	$G(\omega)$ $-\omega_0$ 0
8. $\sin\omega_0 t$	$j\pi\{\delta(\omega+\omega_0) - \delta(\omega-\omega_0)\}$	$jG(\omega)$ $-\omega_0$ 0

10.6 PARSEVAL'S IDENTITY

If $G(\omega) = F\{g(t)\}$, then

$$\int_{-\infty}^{\infty} |g(t)|^2 dt = \frac{1}{2\pi}\int_{-\infty}^{\infty} |G(\omega)|^2 d\omega.$$

In general, if $G(\omega) = F\{g(t)\}$ and $H(\omega) = F\{h(t)\}$, then

$$\int_{-\infty}^{\infty} g(t)h^*(t)dt = \frac{1}{2\pi} \int_{-\infty}^{\infty} G(\omega)H^*(\omega)d\omega ,$$

where * denotes the complex conjugate.

Note: $G(-\omega) = G^*(\omega)$

10.7 CONVOLUTION THEOREM FOR THE FOURIER TRANSFORMS

Let $F(\omega) = F\{f(t)\}$ and $G(\omega) = F\{g(t)\}$. Then the convolution of f and g (i.e., f * g) is defined as

$$f * g = \int_{-\infty}^{\infty} f(\tau)g(t-\tau)d\tau = \frac{1}{2\pi} \int_{-\infty}^{\infty} F(\omega)G(\omega)e^{j\omega t}\, d\omega$$

Hence,

$$\boxed{F\{f * g\} = F(\omega)G(\omega) = F\{f\}\, F\{g\}.}$$

CHAPTER 11

LAPLACE TRANSFORMATION

11.1 DEFINITION OF LAPLACE TRANSFORM

By definition,

$$
L\{g(t)\} = G(s) = \int_0^\infty e^{-st} g(t)dt
$$

where $g(t)$ is a function of the real variable t, and s is a complex variable defined as $s = \delta + j\omega$. The function $g(t)$ is called the original function and the function $G(s)$ is called the image function.

The transformation of a time domain function into a complex frequency domain function is the operation $L\{g(t)\}$.

In order for the Laplace transform to be valid, the following conditions must be satisfied:

1) If the integral in eq.(1) converges for a real $s = s_0$, i.e.,

$$
\lim_{\substack{A \to 0 \\ B \to \infty}} \int_A^B e^{-s_0 t} g(t)dt \quad \text{exists,}
$$

then it converges for all s with $Re(s) > s_0$, and the image function is a single valued analytic function of s in the half-plane $Re(s) > s_0$.

2) g(t) is a piecewise continuous function.

1) It should be noted that in specifying the Laplace transform of a signal, both the algebraic expression and the range of values of s for which this expression is valid is required.

2) The range of values s for which the integral defining the Laplace transform converges is referred to as the region of convergence.

11.2 DEFINITION OF THE INVERSE LAPLACE TRANSFORM

If $L\{g(t)\} = G(s)$, then $g(t) = L^{-1}\{G(s)\}$ is the inverse Laplace transform of $G(s)$. L^{-1} is called the inverse Laplace transform operator.

11.3 COMPLEX INVERSION FORMULA

The inverse Laplace transform of $G(s)$ can be found directly by methods of complex variable theory. The result is

$$g(t) = \frac{1}{2\pi j} \int_{\delta-j\infty}^{\delta+j\infty} e^{st}G(s)ds = \frac{1}{2\pi j} \lim_{T\to\infty} \int_{\delta-jT}^{\delta+jT} e^{st}G(s)ds$$

where δ is chosen such that all the singular points of $G(s)$ lie to the left of the line $\text{Re}(s) = \delta$ in the complex s-plane.

11.4 GENERAL LAPLACE TRANSFORM PAIRS

SIGNAL	TRANSFORM	REGION OF CONVERGENCE
$\delta(t)$	1	All s
$u(t)$	$1/s$	$\text{Re}\{s\} > 0$
$-u(t)$	$1/s$	$\text{Re}\{s\} < 0$
$\dfrac{t^{n-1}}{(n-1)!}\,u(t)$	$1/s^n$	$\text{Re}\{s\} > 0$
$-\dfrac{t^{n-1}}{(n-1)!}\,u(-t)$	$1/s^n$	$\text{Re}\{s\} < 0$
$e^{-\alpha t}\,u(t)$	$\dfrac{1}{s+\alpha}$	$\text{Re}\{s\} > \alpha$
$-e^{-\alpha t}\,u(-t)$	$\dfrac{1}{s+\alpha}$	$\text{Re}\{s\} < -\alpha$
$\dfrac{t^{n-1}}{(n-1)!}\,e^{-\alpha t}\,u(t)$	$\dfrac{1}{(s+\alpha)^n}$	$\text{Re}\{s\} > -\alpha$
$-\dfrac{t^{n-1}}{(n-1)!}\,e^{-\alpha t}\,u(-t)$	$\dfrac{1}{(s+\alpha)^n}$	$\text{Re}\{s\} < -\alpha$
$\delta(t-T)$	e^{-sT}	All s
$[\cos \omega_0 t]\,u(t)$	$\dfrac{s}{s^2+\omega_0^2}$	$\text{Re}\{s\} > 0$
$[\sin \omega_0 t]\,u(t)$	$\dfrac{\omega_0}{s^2+\omega_0^2}$	$\text{Re}\{s\} > 0$
$[e^{-\alpha t}\cos \omega_0 t]\,u(t)$	$\dfrac{s+\alpha}{(s+\alpha)^2+\omega_0^2}$	$\text{Re}\{s\} > -\alpha$
$[e^{-\alpha t}\sin \omega_0 t]\,u(t)$	$\dfrac{\omega_0}{(s+\alpha)^2+\omega_0^2}$	$\text{Re}\{s\} > -\alpha$

11.5 OPERATIONS FOR THE LAPLACE TRANSFORM

1) Linearity of the Laplace Transform:

If $x_1(t) \overset{L}{\longleftrightarrow} X_1(s)$ with region of convergence R_1

and $x_2(t) \longleftrightarrow X_2(s)$ with region of convergence R_2

then

$$ax_1(t) + bx_2(t) \overset{L}{\longleftrightarrow} aX_1(s) + bX_2(s)$$

with region of convergence containing $R_1 \cap R_2$

2) Time Shifting:

If $x(t) \overset{L}{\longleftrightarrow} X(s)$ with region of convergence (ROC) $= R$, then,

$$x(t-t_0) \overset{L}{\longleftrightarrow} e^{-st_0} X(s) \text{ with ROC} = R$$

3) Shifting in the s–Domain:

If $x(t) \overset{L}{\longleftrightarrow} X(s)$ ROC $= R$,

then,

$$e^{s_0 t} x(t) \overset{L}{\longleftrightarrow} X(s-s_0) \text{ with ROC } R_1 = R + Re\{s_0\}$$

Note: The ROC associated with $X(s-s_0)$ is that of $X(s)$, shifted by $Re\{s_0\}$. Thus, for any value s that is in R, the value $s + Re\{s_0\}$ will be in R_1.

For example

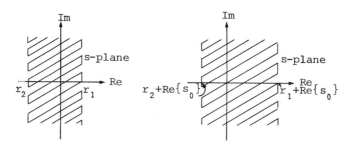

4) Time Scaling:

If $x(t) \overset{L}{\longleftrightarrow} X(s)$ ROC = R,

then

$$x(at) \overset{L}{\longleftrightarrow} \frac{1}{|a|} X\left(\frac{s}{a}\right) \text{ with ROC } R_1 = \frac{R}{a}$$

5) Convolution Property:

If $x(t) \overset{L}{\longleftrightarrow} X_1(s)$ ROC = R_1

(and)

$x_2(t) \overset{L}{\longleftrightarrow} X_2(s)$ ROC = R_2 ,

then

$$x_1(t) * x_2(t) \overset{L}{\longleftrightarrow} X_1(s)X_2(s) \text{ with ROC containing } R_1 \cap R_2$$

6) Differentiation in the Time Domain

If $x(t) \overset{L}{\longleftrightarrow} X(s)$ ROC = R,

then

$$\frac{dx(t)}{dt} \overset{L}{\longleftrightarrow} sX(s) \text{ with ROC containing R}$$

7) Differentiation in the s-domain:

Given:
$$X(s) = \int_{-\infty}^{+\infty} x(t)e^{-st} \, dt$$

Differentiating both sides:
$$\frac{dX(s)}{ds} = \int_{-\infty}^{+\infty} (-t)x(t)e^{-st} \, dt$$

Hence,

$$-t \, x(t) \overset{L}{\longleftrightarrow} \frac{dX(s)}{ds} \qquad ROC = R.$$

8) Integration in the Time Domain:

If $x(t) \overset{L}{\longleftrightarrow} X(s) \qquad ROC = R$,

then

$$\int_{-\infty}^{t} x(\tau)d\tau \overset{L}{\longleftrightarrow} \frac{X(s)}{s} \qquad ROC \text{ contains}$$
$$R \cap \{Re\{s\} > 0\}$$

11.6 HEAVISIDE EXPANSION THEOREM

By the theorem,

$L^{-1}\left\{\dfrac{p(s)}{q(s)}\right\}$, where $q(s)=(s-a_1)(s-a_2)...(s-a_m)$ and

$\qquad\qquad p(s) = $ a polynomial of degree $< m$.

$$= \sum_{n=1}^{m} \frac{p(a_n)}{q'(a_n)} e^{a_n t} \quad \text{(i.e., Heaviside Expansion Formula)}$$

11.7 FINAL AND INITIAL VALUE THEOREM

Initial value theorem:

$$g(0^+) = \lim_{s \to \infty} \{sG(s)\}$$

Final value theorem:

$$g(\infty) = \lim_{s \to 0} sG(s).$$

Note: All poles of sG(s) lie in the left-hand side of the complex s-plane.

CHAPTER 12

TWO PORT NETWORK PARAMETERS

12.1 Z - PARAMETERS

12.1.1 IMPEDANCE OR Z-PARAMETERS

Impedance parameters are defined by the following two sets of equations.

$$v_1 = z_{11}i_1 + z_{12}i_2$$
$$v_2 = z_{21}i_1 + z_{22}i_2$$

where v_1 and v_2 are acting as independent variables.

(Note: A general linear two-port network is being considered.)

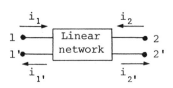

By setting i_1 and i_2 equal to zero, four impedance parameters are defined as follows:

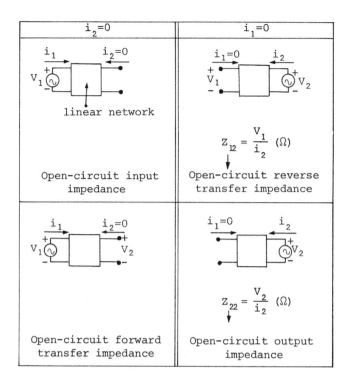

$i_2 = 0$	$i_1 = 0$
linear network	$Z_{12} = \dfrac{V_1}{i_2}$ (Ω)
Open-circuit input impedance	Open-circuit reverse transfer impedance
	$Z_{22} = \dfrac{V_2}{i_2}$ (Ω)
Open-circuit forward transfer impedance	Open-circuit output impedance

(Note: Since i_1 and i_2 are set equal to 0, Z-parameters are also called open-circuit impedance parameters.)

12.2 HYBRID - PARAMETERS

12.2.1 HYBRID OR H-PARAMETERS

The usage of hybrid parameters is for the analysis of transistor circuits. The hybrid parameters are defined by two sets of equations as follows:

$$v_1 = h_{11}I_1 + h_{12}v_2$$

$$I_2 = h_{21}I_1 + h_{22}v_2$$

where v_1 and I_2 are acting as independent variables.

The hybrid parameters are determined by the use of short-circuit and open-circuit conditions.

Hence,

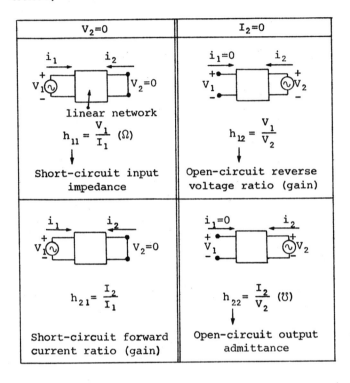

$V_2=0$	$I_2=0$
linear network $h_{11} = \dfrac{V_1}{I_1}$ (Ω) Short-circuit input impedance	$h_{12} = \dfrac{V_1}{V_2}$ Open-circuit reverse voltage ratio (gain)
$h_{21} = \dfrac{I_2}{I_1}$ Short-circuit forward current ratio (gain)	$h_{22} = \dfrac{I_2}{V_2}$ $(\mho)$ Open-circuit output admittance

The hybrid parameters in a two-port network are defined as shown in the network below.

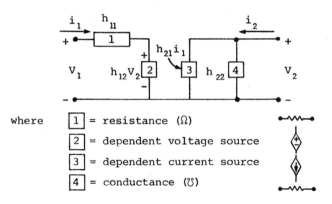

where $\boxed{1}$ = resistance (Ω)

$\boxed{2}$ = dependent voltage source

$\boxed{3}$ = dependent current source

$\boxed{4}$ = conductance $(\mho)$

12.3 ADMITTANCE PARAMETERS

12.3.1 ADMITTANCE OR Y-PARAMETERS

Admittance parameters are described by the following two sets of equations:

$$i_1 = y_{11} v_1 + y_{12} v_2$$

$$i_2 = y_{21} v_1 + y_{22} v_2$$

Then each parameter is defined by setting v_1 or v_2 equal to zero. Hence, Y-parameters are also called the short-circuit admittance parameters.

Thus, by setting:

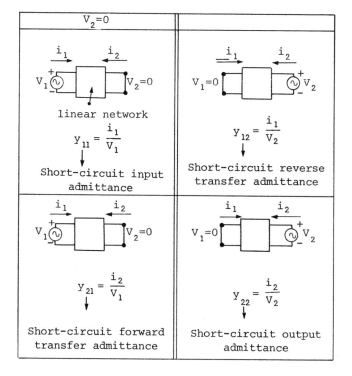

$V_2 = 0$	
linear network $$y_{11} = \frac{i_1}{V_1}$$ Short-circuit input admittance	$$y_{12} = \frac{i_1}{V_2}$$ Short-circuit reverse transfer admittance
$$y_{21} = \frac{i_2}{V_1}$$ Short-circuit forward transfer admittance	$$y_{22} = \frac{i_2}{V_2}$$ Short-circuit output admittance

12.4 Z, Y, AND H PARAMETERS AND RELATIONSHIPS

Let us define the following matrices:

$$[Z] = \begin{bmatrix} z_{11} & z_{12} \\ z_{21} & z_{22} \end{bmatrix}, \quad [Y] = \begin{bmatrix} y_{11} & y_{12} \\ y_{21} & y_{22} \end{bmatrix}, \quad [H] = \begin{bmatrix} h_{11} & h_{12} \\ h_{21} & h_{22} \end{bmatrix}$$

then, $\Delta z = z_{11} z_{22} - z_{12} z_{21}$, $\Delta y = y_{11} y_{22} - y_{12} y_{21}$, $\Delta h = h_{11} h_{22} - h_{12} h_{21}$

are the determinants of $[Z]$, $[Y]$ and $[H]$ respectively.

Now, the conversion between parameters are defined as follows:

(1A) $Z \rightarrow Y \implies \begin{bmatrix} \dfrac{z_{22}}{\Delta z} & \dfrac{-z_{12}}{\Delta z} \\ \dfrac{-z_{21}}{\Delta z} & \dfrac{z_{11}}{\Delta z} \end{bmatrix}$

(1B) $Z \rightarrow H \implies \begin{bmatrix} \dfrac{\Delta z}{z_{22}} & \dfrac{z_{12}}{z_{22}} \\ \dfrac{-z_{21}}{z_{22}} & \dfrac{1}{z_{22}} \end{bmatrix}$

(2A) $Y \rightarrow Z \implies \begin{bmatrix} \dfrac{y_{22}}{\Delta y} & \dfrac{-y_{12}}{\Delta y} \\ \dfrac{-y_{21}}{\Delta y} & \dfrac{y_{11}}{\Delta y} \end{bmatrix}$

(2B) $Y \rightarrow H \implies \begin{bmatrix} \dfrac{1}{y_{11}} & \dfrac{-y_{11}}{y_{11}} \\ \dfrac{y_{21}}{y_{11}} & \dfrac{\Delta y}{y_{11}} \end{bmatrix}$

(3A) $H \rightarrow Z \implies \begin{bmatrix} \dfrac{\Delta h}{h_{22}} & \dfrac{h_{12}}{h_{22}} \\ \dfrac{-h_{21}}{h_{22}} & \dfrac{1}{h_{22}} \end{bmatrix}$

(3B) $H \rightarrow Y \implies \begin{bmatrix} \dfrac{1}{h_{11}} & \dfrac{-h_{12}}{h_{11}} \\ \dfrac{h_{21}}{h_{11}} & \dfrac{\Delta h}{h_{11}} \end{bmatrix}$

CHAPTER 13

DISCRETE SYSTEMS AND Z - TRANSFORMS

13.1 DISCRETE - TIME SYSTEMS

13.1.1 A DISCRETE-TIME SYSTEM

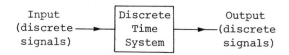

Since discrete signals may be represented by a sequence of numbers, knowing the characteristics of such a sequence is essential.

The characteristics of some general sequences are listed below:

1. Kronecker Delta Sequence:

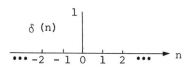

$$\delta(n) = \begin{cases} 1, & n=0 \\ 0, & n=\pm1,\pm2,\ldots \end{cases}$$

$$\delta(n-i) = \begin{cases} 1, & n=i \\ 0, & \text{elsewhere} \end{cases}$$

(Note: i is an arbitrary integer.)

2. Unit Step Sequence:

$$u(n) = \begin{cases} 0, & n=-1,-2, \\ & \quad -3\ldots \\ 1, & n=0,1,2,\ldots \end{cases}$$

3. Unit Alternating Sequence:

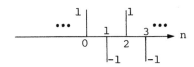

$$u(n) = \begin{cases} 0, & n=-1,-2, \\ & \quad -3\ldots \\ (-1)^n, & n = \\ & 0,1,2,\ldots \end{cases}$$

4. Unit Ramp Sequence:

$$u(n) = \begin{cases} 0, & n=-1,-2,\ldots \\ n, & n=0,1,2,\ldots \end{cases}$$

13.2 FIRST – ORDER LINEAR DISCRETE SYSTEM

A first-order linear discrete system is represented by the linear first-order difference equation as follows:

$$y(n) + A_1 y(n-1) = B_0 u(n) + B_1 u(n-1) \tag{1}$$

where u and y are denoted as the input and output of the system, respectively.

If the input signal is applied at $n = 0$, then eq.(1) becomes

$$y(0) = B_0 u(0) + B_1 u(-1) - A_1 y(-1)$$

where $y(-1)$ is the initial condition of the system.

(Note: $u(-1) = 0$)

In general, where an input signal is applied to $n = j$, then the response of the system is represented as follows:

$$y(j) = B_0 u(j) + B_1 u(j-1) - A_1 y(j-1)$$

103

Since $B_1 u(j-1) = 0$

hence, $y(j) = B_0 u(j) - A_1 y(j-1)$

(Note: The initial condition is defined by $y(j-1)$.)

Summary:

In general, a linear discrete system is described by the relationship as follows:

$$y(n) = B_0 u(n) + B_1 u(n-1) + \ldots + B_i u(n-i) - A_1 y(n-1)$$

$$- A_2 y(n-1) - \ldots - A_N y(n-N)$$

where $\begin{array}{c} B_0 \ldots B_i \\ A_1 \ldots A_N \end{array}$ = constants

and i and N = fixed non-negative integers.

If an input signal is applied at $n = j$, then

$$y(j) = B_0 u(j) - A_1 y(j-1) - A_2 y(j-2) - \ldots - A_N y(j-N)$$

(Note: $u(j-i) = 0$ where $i = 1,2,3,\ldots$)

13.3 CLOSED-FORM IDENTITIES

Closed-form identity is useful in expressing the response of a linear system.

Some generally used closed-form identities are given below:

1) $\displaystyle\sum_{m=0}^{N} r^m = \frac{1 - r^{N+1}}{1 - r}$ where $r \neq 1$

2) $\displaystyle\sum_{m=0}^{N} mr^m = \frac{r}{(1-r)^2} [1 - r^m - mr^m + mr^{m+1}]$ where $r \neq 1$

3) $\displaystyle\sum_{m=0}^{N} m^2 r^m = \frac{r}{(1-r)^3}[(1+r)(1-r^N)-2(1-r)Nr^N-(1-r)^2N^2r^N]$

$$\text{where } r \neq 1$$

13.4 THE Z - TRANSFORM

When a continuous function of time $g(t)$ is sampled at regular intervals of period T, the usual Laplace transform techniques are modified.

The diagrammatic form of a simple sampler together with its associated input-output waveforms is shown below.

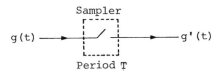

Sampler

$g(t) \longrightarrow \quad \longrightarrow g'(t)$

Period T

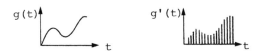

$g(t)$ t $g'(t)$ t

Note: The sampling frequency $\equiv f_s = \dfrac{1}{T}$

Defining the set of impulse function $\delta_T(t)$ by

$$\delta_T(t) \equiv \sum_{n=0}^{\infty} \delta(t-nT)$$

the input-output relationship of the sampler becomes

$$g'(t) = g(t) \cdot \delta_T(t)$$

$$= \sum_{n=0}^{\infty} g(nT) \cdot \delta(t-nT).$$

Note: For a given g(t) and T, the function g'(t) is unique. However the converse is not true.

The variable 'z' is introduced by means of the transformation:

$$z = e^{Ts}$$

and since any function of s can now be replaced by a corresponding function of z, we have

$$G(z) = \sum_{n=0}^{\infty} g(nT) \cdot z^{-n}$$

where $\qquad G'(s) \equiv G(z)$

and $\qquad s = \frac{1}{T} \ln z$

The z operator can now be defined in terms of the Laplace operator by the relationship

$$Z\{g(t)\} \equiv L\{g'(t)\}$$

or

$$Z\{g(t)\} = \Sigma \text{ residues of } \left[\left(\frac{1}{1-e^{Tx}z^{-1}} \right) G(z) \right]$$

The inverse z transform is

$$Z^{-1}\{G(z)\} \equiv g'(t)$$

$$= \frac{1}{2\pi j} \oint G(z) \cdot z^{n-1} dz$$

where the contour of integration encloses all the singularities of the integrand.

106

13.5 PROPERTIES OF Z - TRANSFORM

$g(t)$	$Z\{g(t)\} = G(z)$

1. Linearity: $Af(t)+Bg(t)$

$AF(z)+BG(z)$

2. Left shifting: $g(t+mT)$

$Z^mG(z) - \sum\limits_{r=0}^{m-1} Z^{m-1}g(rT)$

$= Z^mG(z)$ when $g(rT) = 0$,

$0 \le r \le m-1$

3. Right shifting: $g(t-mT)$

$Z^{-m}G(z)$

4. Summation: $\sum\limits_{m=0}^{T/t} g(mT)$

$\left(\dfrac{z}{z-1}\right) G(z)$

5. Differentiating: $tg(t)$

$-Tz \dfrac{d}{dz} G(z)$

6. Integrating: $t^{-1}g(t)$

$-\dfrac{1}{T} \displaystyle\int_0^Z \dfrac{G(z)}{z} dz$

7. Convolution: $\sum\limits_{r=0}^{t} g_1(t-r)g_2(r)$ $G_1(z)G_2(z)$

8. Initial value Theorem

$g(0) = \lim\limits_{|Z| \to \infty} G(z)$

9. Final value Theorem:

$g(\infty) = \lim\limits_{z \to 1} (z-1)G(z)$

if $(z-1)G(z)$ is analytic
for $|z| \ge 1.$

13.6 METHODS OF EVALUATING INVERSE Z – TRANSFORMS

1) Cauchy's residue theorem;

For $t = nT$,

$$g(nT) = \sum_{\text{all } z_k} [\text{residues of } G(z)z^{n-1} \text{ at } z_k]$$

where z_k defines all of the poles of $G(z)z^{n-1}$

2) Partial fractions:

Expand $\dfrac{G(z)}{z}$ into partial fractions. The product of z with each of the partial fractions will then be recognizable from the standard forms in the table of z transforms. Note however that the continuous functions obtained are only valid at the sampling instants.

3) Power series expansion by long division using detached coefficients:

$G(z)$ is expanded into a power series in z^{-1} and the coefficient of the term in z^{-n} is the value of $g(nT)$. i.e., the value of $g(t)$ at the nth sampling instant.

13.6.1 THE Z TRANSFORM AS A MEANS OF DETERMINING APPROXIMATELY THE INVERSE LAPLACE TRANSFORM

Since $Z = e^{Ts}$

$$S^{-1} = \frac{T}{2} \left[\frac{1}{v} - \frac{v}{3} - \frac{4v^3}{45} - \frac{44v^5}{945} - \cdots \right]$$

where

$$v \equiv \frac{1 - z^{-1}}{1 + z^{-1}},$$

the series being very rapid in its convergence. Given $G(s)$ to find its inverse Laplace transform, the following operations are carried out:

1) Divide the numerator and denominator of G(s) by the highest power of s, yielding as an alternate form for G(s) (the quotient of two polynomials in s^{-1}).

2) Choose as a numerical value of T, which makes $2\pi/T$ much larger than the imaginary part of the poles of G(s).

3) Substitute the alternative form for G(s) obtained in (1) above; the expansion for s^{-n} can be determined from the following short table of approximations.

   Do not, at this stage, insert the numerical value for T because tabulations with different intervals may be required.

4) Divide by T.

5) Insert the chosen value for T and divide the numerator by the denominator.

6) The coefficient of z^{-n} is the required value of the function at $t = nT$.

s^{-n}	Z-transform (approximation)
s^{-1}	$\dfrac{T}{2}\left[\dfrac{1 + Z^{-1}}{1 - Z^{-1}}\right]$
s^{-2}	$\dfrac{T^2}{12}\left[\dfrac{1 + 10Z^{-1} + Z^{-2}}{(1 - Z^{-1})^2}\right]$
s^{-3}	$\dfrac{T^3}{3}\left[\dfrac{Z^{-1} + Z^{-2}}{(1 - Z^{-1})^3}\right]$
s^{-4}	$\dfrac{T^4}{144}\left[\dfrac{1 + 20Z^{-1} + 102Z^{-2} + 20Z^{-3} + Z^{-4}}{(1 - Z^{-1})^4}\right]$
s^{-5}	$\dfrac{T^5}{124}\left[\dfrac{Z^{-1} + 11Z^{-2} + 11Z^{-3} + Z^{-4}}{(1 - Z^{-1})^4}\right]$
s^{-6}	$\dfrac{T^6}{4}\left[\dfrac{Z^{-2} + 2Z^{-3} + Z^{-4}}{(1 - Z^{-1})^6}\right]$
s^{-7}	$\dfrac{T^7}{8}\left[\dfrac{Z^{-2} + 3Z^{-3} + 3Z^{-4} + Z^{-5}}{(1 - Z^{-1})^7}\right]$

13.7 Z - TRANSFORM PAIRS

Transform Pair Signal	Transform	ROC
$\delta[n]$	1	All z
$u[n]$	$\dfrac{1}{1 - Z^{-1}}$	$|Z| > 1$
$u[-n - 1]$	$\dfrac{1}{1 - Z^{-1}}$	$|Z| < 1$
$\delta[n - m]$	Z^{-m}	All Z except 0 (if m > 0) or ∞ (if m < 0)
$\alpha^n u[n]$	$\dfrac{1}{1 - \alpha Z^{-1}}$	$|Z| > |\alpha|$
$-\alpha^n u[-n - 1]$	$\dfrac{1}{1 - \alpha Z^{-1}}$	$|Z| < |\alpha|$
$n\alpha^n u[n]$	$\dfrac{\alpha Z^{-1}}{1 - \alpha Z^{-1})^2}$	$|Z| > |\alpha|$
$-n\alpha^n u[-n - 1]$	$\dfrac{\alpha Z^{-1}}{(1 - \alpha Z^{-1})^2}$	$|Z| < |\alpha|$
$[\cos \omega_0 n] \, u[n]$	$\dfrac{1 - [\cos \omega_0] Z^{-1}}{1 - [2\cos \omega_0] Z^{-1} + Z^{-2}}$	$|Z| > 1$
$[\sin \omega_0 n] \, u[n]$	$\dfrac{[\sin \omega_0] Z^{-1}}{1 - [2\cos \omega_0] Z^{-1} + Z^{-2}}$	$|Z| > 1$

CHAPTER 14

TOPOLOGICAL ANALYSIS

14.1 INCIDENT MATRIX

By definition, an augmented incident matrix Aa, is an n(nodes) × b(branches) matrix of a directed graph of any planar network, i.e.,

$$Aa = [a_{ij}]_{n \times b}$$

where

$$a_{ij} = \begin{cases} 1 & \text{when branch bj is incident to node } n_i \text{ and the reference current, } i_j, \text{ leaves the node.} \\ -1 & \text{when branch bj is incident to node } n_j \text{ and the reference current, } i_j, \text{ enters the node.} \\ 0 & \text{when branch bj is not incident to node } n_i. \end{cases}$$

The incident matrix Aa can be represented as:

$$Aa = \begin{matrix} & b_1 \ b_2 \ b_3 \ \dots \ b_j \\ \begin{matrix} n_1 \\ n_2 \\ n_3 \\ \vdots \\ n_i \end{matrix} & \begin{bmatrix} & & & & \\ & & & & \\ & & & & \\ & & & & \\ & & & & \end{bmatrix} \end{matrix}$$

Note: An incidence submatrix can be obtained by taking out any one of the rows of the incidence matrix Aa.

14.2 THE CIRCUIT (LOOP) MATRIX

By definition, an augmented circuit matrix, B_a, is an $\ell \times b$ matrix where ℓ = loops and b = branches.

Hence, $Ba = [b_{ij}]_{\ell \times b}$

where

$$
b_{ij} = \begin{cases} 1 & \text{when branch bj is in loop } \ell_i \text{ and is} \\ & \text{oriented in the same direction.} \\ -1 & \text{when branch bj is in loop } \ell_i \text{ and} \\ & \text{is oriented in the opposite direction.} \\ 0 & \text{when branch bj is not in loop } \ell_i. \end{cases}
$$

The circuit matrix can be represented as:

$$
Ba = \begin{array}{c} \\ \ell_1 \\ \ell_2 \\ \ell_3 \\ \vdots \\ \ell_i \end{array} \begin{array}{c} b_1 b_2 b_3 \ldots b_j \\ \left[\begin{array}{cccc} & & & \\ & & & \\ & & & \\ & & & \\ & & & \end{array} \right] \end{array}
$$

14.3 FUNDAMENTAL (LOOP) MATRIX

The fundamental loop (circuit) matrix, B_f, is defined as a $[b-(n-1)] \times b$ matrix where b = branches and n = nodes. i.e.,

$$B_f = [b_{ij}]_{[b-(n-1)] \times b}$$

where

$$b_{ij} = \begin{cases} 1 & \text{when branch bj is in the fundamental loop } \ell_i \text{ and is oriented in the same direction.} \\ -1 & \text{when branch bj is in the fundamental loop } \ell_i \text{ and is oriented in the opposite direction.} \\ 0 & \text{when branch bj is not in the fundamental loop } \ell_i. \end{cases}$$

Note: A fundamental loop cannot contain more than one chord; a chord is any branch of a cotree and a cotree is the set of all branches not in a tree. (A tree is defined in Chapter 5.)

The matrix representation of B_f is as follows:

$$B_f = \begin{array}{c} \\ \ell_1 \\ \ell_2 \\ \ell_3 \\ \vdots \\ \ell_i \\ \vdots \\ \ell_{b-(n-1)} \end{array} \overset{\displaystyle b_1\ b_2\ b_3\ \ldots\ b_j \ldots\ b_b}{\left[\begin{array}{ccccccc} & & & & & & \\ & & & & & & \\ & & & & & & \\ & & & & & & \\ & & & & & & \\ & & & & & & \\ & & & & & & \end{array} \right]}$$

CHAPTER 15

NUMERICAL METHODS

15.1 NEWTON'S METHOD

Newton's method is used to find the roots of a polynomial equation.

Consider the equation:

$$F(s) = k_4 s^4 + k_3 s^3 + k_2 s^2 + k_1 s^1 + k_0 = 0.$$

In order to determine the real value of $s = s'$ such that $F(s) = 0$, the Newton's method is used as follows:

Since all the coefficients of $F(s)$ are of the same sign and real, the complex roots of $F(s)$ are complex conjugate pairs and the real roots, if any, are negative.

Now, by inspection, we begin with a guess, $s = s_0$, for the root. Also let $s_1 = s_0 - h_0$ where

s_1 = a closer approximation of the root obtained from s_0 and

$$h_0 = s_0 - s_1 = \frac{-F(s_0)}{F'(s_0)} \quad \text{and} \quad F'(s_0) = \frac{d}{ds} F(s) \bigg|_{s=s_0}$$

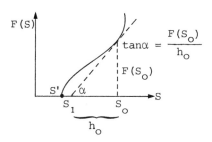

Then, in general,

$$S_{i+1} = S_i - \frac{F(s_i)}{F'(s_i)} = (i+1) = \text{iteration of the initial approximation } s_0.$$

i.e., S_i = the previous approximation (or guess) of the root.

and S_{i+1} = the new approximation of $F'(S_i)$, which is the derivative of $F(S)$ evaluated at $S = S_i$.

Finally, the iteration is stopped when S_{i+1} is approximately equal to S_i, indicating that the value of S_i is the root s'.

15.2 SIMPSON'S RULE

Simpson's rule states that

$$\int_{x_0 = a}^{x=b} y(x)dx \cong \frac{h}{3}(y_0 + 4y_1 + 2y_2 + 4y_3 + 2y_4 + \ldots + 2y_{n-2}$$
$$+ 4y_{n-1} + y_n)$$

where $h = \frac{b-a}{n}$ (note: n is even.)

and $y_0 = y(a)$, $y_1 = y(a+h)$, $y_2 = y(a+2h)$, $y_n = y(a+nb) = y(b)$

Note: The more values between the limits of integration taken, (the larger n is), the more accurate the result will be.

Simpson's rule is a simple and reasonably accurate method which can be used to write programs for digital computers or programmable calculators. The flow chart shown below is for Simpson's rule of integration.